Freezing Time and Unraveling the Cosmos

Exploring Cryogenics, Time Travel, and M Theory

Alfonso Borello

Villaggio Publishing Ltd

CONTENTS

BEFORE THE INTRODUCTION

Before delving into the introduction, it's pivotal to acknowledge the evolving landscape of M theory, a theoretical framework that continues to navigate through the turbulent waters of scientific inquiry. Despite its relative youthfulness, M theory confronts formidable obstacles that stem from its intricate mathematical formulations, which pose challenges in terms of testability and validation. Furthermore, the absence of tangible experimental evidence corroborating its postulates, particularly regarding the existence of extra dimensions and other conjectures, imbues the theory with an aura of conjecture and speculation. However, within the depths of these challenges lie tantalizing prospects and profound rewards awaiting discovery. Even in the absence of definitive empirical verification, the foundational tenets of M theory hold profound implications, extending far beyond the realm of theoretical physics. Consider the parallels that can be drawn between the principles of multi-dimensional thought inherent in M theory and the complexities encountered in everyday life scenarios.

Picture yourself grappling with the intricacies of managing a diverse and dynamic team within a professional setting. By adopting a multi-dimensional perspective, one can unravel hidden dynamics and cultural nuances that may clandestinely influence team dynamics and performance. Embracing the complexity inherent in such scenarios allows for the deconstruction of daunting challenges into more digestible components, all while retaining sight of the overarching objectives. Moreover, fostering an environment that embraces diverse perspectives enriches the problem-solving process, offering insights and solutions that transcend conventional wisdom.

Indeed, the principles espoused by M theory serve as more than mere conjectures awaiting empirical validation. They represent a paradigm shift in cognitive frameworks, offering invaluable tools for navigating the labyrinthine corridors of complexity that characterize both the scientific frontier and the myriad challenges encountered in everyday life. By embracing multi-dimensional thinking, individuals are empowered to confront and conquer seemingly insurmountable obstacles, leveraging a holistic approach that transcends conventional boundaries.

So, as we stand at the precipice of discovery, poised to embark on a journey that may lead us to unlock the enigmatic secrets of the universe, let us heed the call to adventure with unwavering resolve. For even if the answers elude us within the confines of our current understanding, the pursuit of knowledge and the expansion of our cognitive horizons herald a quest of infinite possibilities.

A BRIEF HISTORY OF SCIENTIFIC THEORIES

The history of scientific theories is a fascinating journey through human curiosity, ingenuity, and discovery. Scientific theories are the backbone of our understanding of the natural world, providing frameworks for explaining observed phenomena, making predictions, and guiding further research. Here's why the history of scientific theories matters:

1. Understanding Progression: Studying the history of scientific theories allows us to trace the evolution of ideas over time. We can see how earlier theories laid the groundwork for later developments, leading to our current understanding. For example, the geocentric model of the universe proposed by ancient astronomers eventually gave way to the heliocentric model through the work of scientists like Copernicus and Galileo.

2. Contextualizing Discoveries: Scientific theories don't emerge in a vacuum; they are shaped by the cultural, social, and intellectual context of their time. By examining the historical context in which theories arose, we gain insights into the motivations, biases, and constraints that influenced scientific thinking. For instance, Darwin's theory of evolution by natural selection was deeply influenced by the scientific, religious, and philosophical debates of the 19th century.

3. Learning from Mistakes: Not all scientific theories stand the test of time. Some are refined, revised, or even replaced as new

evidence comes to light. Understanding the history of theories allows us to appreciate the iterative nature of science and the importance of skepticism, critical thinking, and empirical evidence in advancing knowledge. For example, the phlogiston theory of combustion was eventually replaced by the modern understanding of oxidation through the work of scientists like Lavoisier.

4. Inspiring Innovation: Many scientific breakthroughs are inspired by previous theories or attempts to solve unanswered questions. By studying the history of scientific theories, researchers can uncover overlooked ideas, alternative approaches, or unresolved puzzles that may spark new avenues of inquiry. For instance, Einstein's theory of general relativity was influenced by his study of Maxwell's equations and the failures of classical mechanics to explain certain phenomena.

5. Appreciating Diversity: Scientific theories often coexist and compete with one another, reflecting the diversity of perspectives, methodologies, and interpretations within the scientific community. By exploring the history of theories, we gain a deeper appreciation for the richness and complexity of scientific thought, as well as the ongoing dialogue and debate that drives progress. For example, different interpretations of quantum mechanics, such as the Copenhagen interpretation and the Many-Worlds interpretation, continue to be topics of lively discussion among physicists.

Indeed, the history of scientific theories offers valuable insights into the process of discovery, the dynamics of scientific communities, and the nature of knowledge itself. By studying this history, we can better appreciate the achievements of past scientists, critically evaluate current theories, and anticipate the directions of future research.

EMBARKING ON A JOURNEY INTO THE MULTIVERSE: UNVEILING THE DEPTHS OF M THEORY

Hey there, fellow cosmic adventurers! Let's dive into the mind-bending world of M theory, where our universe isn't just the four dimensions we're used to but a mind-blowing eleven-dimensional playground of possibilities

Picture this: Beyond the everyday world we see and feel lies a hidden realm, where space and time twist and dance in ways we can barely fathom. That's where M theory comes in, trying to wrap its head around the crazy intricacies of the cosmos and give us a peek behind the curtain of reality.

So, how did we end up on this wild ride? Well, back in the '90s, physicists were scratching their heads over these things called superstring theories. They suggested that instead of tiny points, fundamental particles are like tiny vibrating strings, each humming its own unique tune. But there were holes in these theories, gaps that needed filling.

Enter M theory, our hero in this cosmic saga. It swoops in with the bold idea that our universe isn't just four dimensions—oh no, it's got a whopping eleven! But here's the kicker: most of these dimensions are curled up so tiny, like the coils of a spring, that we can't even see them with our most powerful microscopes. Talk about hiding in plain sight!

And that's not all! M theory throws another curveball with something called "branes." Imagine sheets of energy vibrating in this multidimensional space, kind of like the strings but on a whole new level. Our universe? It's like a slice of cosmic pizza floating in this eleven-dimensional soup.

Now, hold onto your hats because here's where it gets really trippy. M theory suggests that our universe isn't alone in this cosmic dance. Nope, there could be a whole bunch of other universes out there, each with its own set of rules and constants. It's like a multiverse party, and we're just one guest among many!

So, what's the big deal with all this? Well, imagine if M theory turns out to be the ultimate cheat code for understanding the universe. We're talking about finally nailing down the answers to some of the biggest questions in science, from why stuff falls down to what's really going on with all that dark matter and energy stuff.

But hey, it's not all sunshine and rainbows. M theory's still in its infancy, struggling with some pretty hefty challenges. The math alone is enough to make your brain hurt, and don't even get me started on trying to find evidence for all those extra dimensions!

Still, we're not giving up hope. Scientists are pushing the limits, tweaking the theory, and dreaming up new ways to put it to the test. Who knows? Maybe one day we'll crack the cosmic code wide open and unlock the secrets of the universe.

So, fellow cosmic explorers, are you ready to strap in and join us on this wild ride into the unknown? Because with M theory, the adventure is just getting started!

EXPLORING THE COSMIC ENIGMA: THE INTRIGUING REALM OF PARALLEL UNIVERSES

Have you ever pondered the possibility of living a slightly different life, right now, in another world? Such a thought may seem like the stuff of science fiction, but experts suggest that beyond the confines of our known universe lie realms where such scenarios may not only be possible but commonplace. Welcome to the tantalizing domain of parallel universes, where the boundaries of reality blur and the imagination knows no bounds.

Unveiling Parallel Universes

Experts believe that the cosmos holds more than meets the eye. The notion of parallel universes, often discussed in scientific circles, proposes the existence of alternate realities existing alongside our own. These universes, if they indeed exist, could manifest in various forms, with at least four types theorized to exist. One type, in particular, is believed to be in close proximity to our own, occupying the same space yet remaining invisible and inaccessible due to their unique properties.

String Theory And The Fabric Of Reality

At the heart of the parallel universe hypothesis lies String Theory, a groundbreaking concept that suggests the fundamental building blocks of the universe are not solid points but tiny vibrating strings. These strings, vibrating in different ways, give rise to the matter and energy that populate our world. Furthermore, String Theory posits the existence of extra dimensions beyond the familiar four, offering the possibility of hidden realms that coexist with our own.

Exploring Multiverses And Cosmic Inflation

Within the framework of String Theory, the concept of multiverses emerges, presenting a dizzying array of potential realities. These multiverses, if they exist, could encompass a myriad of universes, each with its own set of physical laws and constants. The idea of cosmic inflation, a theory describing the rapid expansion of the universe in its infancy, provides further fuel to the notion of a multiverse teeming with diverse possibilities.

The Shape Of The Universe And Bubble Nucleation

The shape of the universe itself remains a subject of intense debate among physicists. Some propose it resembles a giant sphere, while others liken it to a flat expanse or even a series of interconnected bubbles. The concept of bubble nucleation suggests that these bubbles, formed within a larger universe, could give rise to entirely new universes, each with its own distinct characteristics and evolutionary path.

Challenges And Controversies

Despite the allure of parallel universes, skeptics remain unconvinced of their existence. The proliferation of theories and the rebranding of concepts have blurred the lines between physics and philosophy, leading to heated debates within the scientific community. Furthermore, the complexity of String Theory and the inability to verify its predictions in laboratory experiments pose significant challenges to its acceptance as a comprehensive theory of the universe.

The Quest For A Unified Theory

Amidst the uncertainty and speculation, scientists continue their quest for a unified theory that can reconcile the fundamental forces of the universe. While String Theory offered tantalizing insights, its evolution into M-Theory, or the Theory of Membranes, presents a more comprehensive framework that operates in eleven dimensions. M-Theory posits the existence of membranes, or "branes," on which our reality may be imprinted, offering a potential solution to the mysteries of parallel universes. As we navigate the cosmic maze of parallel universes, we are confronted with profound questions about the nature of reality and our place within it. Could alternate versions of ourselves exist in distant dimensions, living out lives vastly different from our own? Are we but one among countless worlds scattered across the multiverse? While the answers remain elusive, the journey into the unknown continues, fueled by curiosity, imagination, and the relentless pursuit of truth.

EXPANDING THE IDEA OF WHITE HOLES: BEYOND ONE-WAY DOORS

White holes, the theoretical cosmic counterparts to black holes, are indeed fascinating and warrant further exploration. Here are some ways to expand on their concept:

1. Beyond One-Way Doors

While often portrayed as "time-reversed black holes," white holes may not simply regurgitate everything swallowed by a black hole. They could be sources of entirely new matter and energy, distinct from what entered their past counterparts.

Instead of a one-way door, imagine them as cosmic springs, spewing forth exotic matter and energy from another part of the universe, or perhaps even a different dimension.

2. Properties And Effects

Unlike black holes with their immense gravity, white holes are speculated to have repulsive gravity, pushing anything near them away. This makes them even harder to observe or interact with.

Theories propose that they might be incredibly bright, emitting vast amounts of radiation across all spectrums, potentially illuminating vast regions of space.

Their existence could have significant implications for the universe's overall energy balance and expansion, potentially influencing the fate of our own universe.

3. Connection To Other Theories

White holes could be linked to other cosmological concepts like wormholes, serving as the exit point for matter traveling through these hypothetical shortcuts across spacetime.

Some interpretations of String Theory suggest that white holes might be singularities at the "birth" of new universes, spewing forth matter and energy that eventually coalesce into new cosmoses.

4. The Search For White Holes

With their repulsive gravity and potential invisibility, directly detecting white holes is incredibly difficult. However, astronomers might be able to indirectly infer their presence by studying their effects on surrounding matter or radiation patterns.

Future advancements in telescopes and gravitational wave detectors could increase our chances of identifying these elusive objects, potentially unlocking crucial insights into the universe's structure and evolution.

5. Beyond Science Fiction

White holes have captured the imaginations of science fiction writers and filmmakers, often depicted as gateways to other universes or sources of unimaginable power. While these portrayals are usually fictional, they highlight the potential these theoretical entities hold for sparking curiosity and inspiring further scientific exploration.

Remember, white holes are still just theoretical, and their true nature remains shrouded in mystery. But by delving deeper into their properties and potential connections to other cosmological phenomena, we can continue to push the boundaries of our understanding of the universe and its secrets.

IS THERE ANYTHING OUT THERE THAT SUGGESTS THAT AN ANTI-GRAVITATIONAL FIELD WOULD CREATE SUCH EFFECT?

While white holes are often presented as having "negative gravity" or a repulsive force, it's crucial to differentiate this from a true "anti-gravitational field." Here's why:

Gravity's Nature

- Gravity, as described by General Relativity, isn't a force in the traditional sense. It's instead an emergent phenomenon arising from the curvature of spacetime caused by mass and energy.

- Imagine placing a bowling ball on a trampoline. It creates a dip in the fabric, and smaller objects placed nearby naturally roll towards the dip, not because of a "pulling" force, but due to the curved path dictated by the dip.

White Holes And Repulsion

- White holes, if they exist, wouldn't create an opposing force like a true "anti-gravity" field. Instead, their hypothesized repulsive nature stems from the way they might expel matter and energy at incredibly high speeds due to their internal pressure.

- Think of a garden hose: water exiting under pressure pushes outwards, not because of a negative pressure but due to the inherent energy pushing it out.

Theories Around Expulsion

- Different theories have varying explanations for what white holes might expel. Some suggest exotic particles or radiation, while others propose entirely new forms of matter.

- No current theory suggests they directly "suck in" matter from somewhere else in the universe, unlike black holes and their suspected Hawking radiation.

Current Understanding

- While intriguing, white holes remain purely theoretical, and concrete evidence for their existence is lacking. Our understanding of gravity and fundamental physics still doesn't offer a definitive explanation for how they might function.

Open Questions

- Could their existence shed light on singularities at the Big Bang or even the birth of new universes?

- How would their "expulsion" interact with matter and energy in its surroundings?

Exploring these questions motivates further research into gravity, exotic matter, and the universe's fundamental structure.

While true "anti-gravity" might remain elusive, the mysteries surrounding white holes offer a fascinating avenue for scientific exploration and discovery.

COULD I ENTER A BLACK HOLE AND GET OUT OF A WHITE HOLE IN A REGION OF THE UNIVERSE WAY DIFFERENT IN SPACE AND TIME FROM WHERE I STARTED?

Unfortunately, based on our current understanding of physics, the answer is a resounding no. Here's why:

Black Holes And The Singularity

Black holes are incredibly dense objects with such immense gravity that nothing, not even light, can escape their pull once it crosses a specific point called the event horizon.

Within a black hole, gravity becomes infinitely strong as it approaches a point of singularity, where all the mass is concentrated in an infinitely small space.

Our current understanding of physics breaks down at the singularity, and we don't know what happens to matter or information that falls in. There's no guarantee, nor any physical theory suggesting, that anything entering a black hole would survive or emerge elsewhere.

White Holes: Theoretical And Mysterious

White holes are theoretical entities postulated as the counterparts to black holes, spewing out matter and energy instead of sucking it in. However, their existence remains unproven, and their properties are largely speculative.

Even if white holes existed, we have no evidence or physical theory suggesting they act as exits from black holes. They are likely independent objects, not connected to black holes in any way.

Space And Time Travel

The passage of time within a black hole is theorized to be extremely warped due to its intense gravity. From an external observer's perspective, time would appear to slow down infinitely as an object approaches the event horizon.

While some science fiction concepts explore traversing space and time through black holes, these remain purely fictional due to the physical limitations mentioned above.

Further Explorations

Understanding black holes and the nature of gravity at the quantum level continues to be a significant challenge in physics. Future advancements in our understanding of these phenomena might reveal new possibilities, but currently, the journey through a black hole is a one-way trip with no known return.

Even if white holes do exist, their hypothetical properties wouldn't offer a mechanism for space and time travel as depicted in fiction.

While the notion of traveling through black holes and emerging into different universes is captivating, it remains outside the realm of current scientific understanding. The mysteries surrounding black holes and white holes, however, continue to inspire scientific exploration and push the boundaries of our knowledge about the universe.

PERHAPS A BLACK HOLE IS JUST A BOTTOMLESS PIT WHERE MATTER ENTERS AND ACCUMULATES.

It's an interesting and understandable way to think about black holes, but based on our current understanding of physics, it's not entirely accurate. Here's why:

Black Holes Aren't Bottomless

While the gravity of a black hole is incredibly strong, it's not infinitely strong. There's a specific limit, called the Schwarzschild radius, beyond which anything, even light, cannot escape. So, it's not just a deep pit with no end; it has a very real boundary.

Beyond the Event Horizon

Once something crosses the event horizon, we can't directly observe it anymore. However, it doesn't simply fall into an empty void. The immense gravity compresses it to an unimaginably small point called a singularity.

The Mystery Of The Singularity

Here's where our knowledge gets shaky. At the singularity, gravity becomes infinitely strong, and our current theories of physics break down. We don't know what happens to matter or information there. It could be destroyed, compressed into some exotic form, or even potentially lead to other dimensions or universes (though this is highly speculative).

Accumulation And Hawking Radiation

Black holes do "accumulate" matter in the sense that anything that crosses the event horizon is essentially trapped inside. However, they're not passive absorbers. According to Stephen Hawking's theory, black holes actually emit a faint radiation called Hawking radiation, slowly losing some of their mass and energy over vast timescales.

So, What's The Takeaway?

Black holes are complex and fascinating objects, and while we can't directly observe what happens inside them, we have a deeper understanding of their properties and limitations than simply "bottomless pits." The mysteries surrounding them continue to fuel research and challenge our understanding of gravity and the universe's fundamental laws.

COULD A ONE WAY TUNNEL, IF WELL DESIGNED, TRANSPORT PASSENGERS IN DIFFERENT LOCATIONS IN TIME OR EVEN INTO SPACE?

Unfortunately, with our current understanding of physics, building a tunnel that transports passengers through time or space isn't feasible. Here's why:

Space Travel

Distance: The universe is vast. Traveling to even the nearest star Proxima Centauri would take thousands of years with our current technology. Even the most advanced tunnel wouldn't be able to overcome these immense distances without violating known physical laws.

Speed Limits: The speed of light is a fundamental limit in the universe. Tunnels wouldn't be able to bypass this limit, which means travel times would still be significant and potentially impractical for most purposes.

Time Travel

Causality: Traveling backwards in time presents a major paradox. Changing the past could have unintended consequences on the present, potentially creating a contradictory or unstable timeline. Physics currently has no known mechanisms to reconcile this contradiction.

Energy and Relativity: Bending the fabric of spacetime to travel through time requires immense amounts of energy, potentially exceeding what we can generate or control with current technology. Additionally, General Relativity suggests that time dilation occurs naturally near massive objects or at high speeds, but this wouldn't be the same as true time travel.

Alternative Concepts

While physical tunnels for time or space travel seem impossible with our current knowledge, there are some theoretical ideas worth exploring:

Wormholes: These hypothetical shortcuts through spacetime could connect distant regions in space, but traversing them might involve overcoming similar limitations as traditional space travel.

Cryogenics: While not technically time travel, suspending individuals in cryogenic sleep could theoretically allow them to "travel" into the future by slowing down their biological processes.

The Future

Our understanding of physics continues to evolve, and future discoveries might offer new possibilities for traveling across space or time. However, based on what we know today, building a physical tunnel for such purposes remains firmly in the realm of science fiction.

Remember, even though these concepts might seem fantastical, they serve as thought experiments that push the boundaries of our current understanding and inspire further scientific exploration.

DELVING DEEPER INTO CRYOGENICS: FREEZING TIME OR SCI-FI FANTASY?

Cryogenics, the practice of preserving life through extreme cold, is a captivating yet controversial topic. While often portrayed as a solution to mortality in science fiction, the reality is much more nuanced. Let's dive into the science, ethical concerns, and future prospects of this intriguing field:

1. The Science Behind Cryonics

- The heart of cryogenics lies in vitrification, a process where biological materials are cooled rapidly enough to solidify into a glass-like state without ice formation. This aims to prevent cellular damage caused by ice crystals during traditional freezing.

- Once vitrified, tissues are stored at extremely low temperatures (-196°C or -321°F), hoping to halt biological processes and essentially "pause" biological time.

- The ultimate goal is to revive individuals in the future when technology advances enough to repair any cellular damage and restore life functions.

2. Ethical And Practical Challenges

- Despite significant research, achieving successful revival after cryopreservation remains a major challenge. The complexity of biological systems and the potential for irreversible damage during the freezing and thawing processes are significant hurdles.

- Ethical concerns surround informed consent, the long-term viability of cryopreservation facilities, and the potential social and legal implications of reviving individuals in a vastly different future.

- The financial costs associated with cryopreservation are substantial, further limiting accessibility and raising questions about its ethical distribution.

3. Future Prospects

- While the scientific and ethical challenges are daunting, advancements in fields like regenerative medicine and nanotechnology offer potential solutions for cellular repair and tissue reconstruction in the future.

- Research continues to refine cryopreservation techniques and improve understanding of cellular damage mechanisms.

- Openness and transparency are crucial for building public trust and addressing ethical concerns surrounding cryonics.

4. Beyond Freezing Humans

- Cryogenics has applications beyond human preservation. Cryopreserving tissues and organs for transplantation, preserving genetic material for endangered species, and developing advanced bioprinting techniques are some promising areas of research.

TECHNICALITIES

Cryonics, as a field, relies on intricate techniques and advanced technologies to achieve the preservation of biological tissues at ultra-low temperatures. Let's delve deeper into the specific methods employed in cryopreservation:

1. Cryoprotectants

- Cryoprotectants are chemical compounds designed to prevent ice crystal formation within biological tissues during freezing.
- These substances penetrate the cells and tissues, replacing water molecules and reducing the formation of ice crystals that can cause cellular damage.
- Common cryoprotectants include glycerol, dimethyl sulfoxide (DMSO), ethylene glycol, and various types of sugars.

2. Cooling Methods

- Rapid cooling is crucial in cryopreservation to prevent ice crystal formation and minimize cellular damage.
- Controlled-rate freezing: This method involves gradually lowering the temperature of the biological sample at a controlled rate to allow the cryoprotectants to penetrate the tissues effectively.
- Vitrification: In this process, tissues are cooled rapidly to extremely low temperatures, typically below -130°C (-202°F),

causing them to solidify into a glass-like state without the formation of ice crystals. Vitrification minimizes cellular damage and preserves the structural integrity of the tissues.

- Ultra-rapid cooling: Some advanced cryopreservation techniques utilize ultra-rapid cooling methods, such as plunge freezing in liquid nitrogen or direct immersion in cryogenic liquids, to achieve even faster cooling rates and enhance tissue preservation.

3. Perfusion Techniques

- Perfusion involves the circulation of cryoprotectant solutions through the blood vessels of the biological tissue to ensure uniform distribution and penetration of the cryoprotectants.
- This method is commonly used in whole-body cryopreservation to deliver cryoprotectants throughout the body before cooling.

4. Cryogenic Storage

- After the cryopreservation process, the biological samples are transferred to specialized storage containers, such as cryogenic dewars or tanks, filled with liquid nitrogen.
- Liquid nitrogen maintains temperatures below -196°C (-321°F), ensuring long-term preservation of the biological samples without significant deterioration.

5. Monitoring And Quality Control

- Cryopreservation procedures require careful monitoring of temperature, pressure, and other environmental conditions to ensure the optimal preservation of biological tissues.
- Quality control measures, such as periodic inspections

and maintenance of cryogenic storage facilities, are essential to minimize the risk of temperature fluctuations and equipment failures that could compromise the integrity of the samples.

Indeed, with the use of these sophisticated techniques and technologies, cryonics practitioners aim to achieve the long-term preservation of biological tissues, with the ultimate goal of potentially reviving individuals in the future when medical science and technology have advanced sufficiently.

TIME MACHINE TALES: FANTASTICAL SCENARIOS FOR YOUR FIRST TEMPORAL ADVENTURE!

Ah, let's leave the thought-provoking idea of taking a vacation at the event horizon for a moment and jump back into the enthralling world of time travel! But hey, before we get too carried away, let's remember that time machines are still firmly in the realm of science fiction—for now, at least! So, any discussion about what happens when you flick the switch on one of these babies is purely speculative, but boy, is it fun to speculate! Here are a few wild scenarios to get your creative gears turning:

1. The Big Bang Blastoff

Picture this: flicking on the time machine isn't just a push of a button; it's a symphony of calibrations, energy surges, and calculations. Then, instead of an instant zap through time, you feel a gradual acceleration, like riding a cosmic roller coaster. As the machine revs up, you're treated to a mind-bending fast-forward through the universe's history, from the explosive birth of the cosmos to the present day, all compressed into a whirlwind of sights and sounds.

2. Glitches And Paradoxes Galore

Now, let's spice things up with a bit of chaos! What if, instead of a smooth ride, the time machine coughs and splutters upon activation? Suddenly, you're careening through time like a pinball, bouncing between random points in history with no rhyme or reason. Or, even spookier, maybe firing up the machine sets off a chain reaction of temporal paradoxes, sending shockwaves through the fabric of spacetime and causing all sorts of unpredictable havoc.

3. A Date With Destiny

But hey, let's not discount the possibility of precision! Maybe this time machine is a well-oiled temporal marvel, offering pinpoint accuracy in its travels. Flip the switch, and voila! You're whisked away to your chosen destination, whether it's a roaring twenties speakeasy or a distant future where robots rule the world. Talk about a tailor-made adventure!

4. Mind Over Matter

What if time travel isn't about physical journeys but rather a trip for the mind? Imagine flipping the switch and feeling your consciousness untethered from your body, soaring through the annals of history or skipping ahead to peek at the future. It's like watching a blockbuster movie, but you're not just a passive observer—you're right there in the thick of it, soaking up every moment.

5. The Grab Bag Of Temporal Thrills

Of course, who's to say it can't be a bit of everything? Maybe this time machine is a quirky blend of chaos and control, offering a tantalizing mix of thrilling adventures and head-scratching mysteries. Each flick of the switch is a roll of the dice, with the potential for heart-pounding excitement, unexpected detours, and maybe even a few mind-bending paradoxes thrown in for good measure.

But hey, let's not get too lost in the what-ifs. While we wait for science to catch up with our wildest dreams, let's keep our imaginations fired up and ready for whatever time-traveling adventures the future might hold!

Q&A

Q: What is String Theory and how does it relate to M Theory?

A: String Theory is a theoretical framework in physics that posits that the fundamental building blocks of the universe are not particles, but rather tiny, vibrating strings. M Theory is an extension of String Theory that proposes the existence of multiple dimensions beyond the familiar four (three spatial dimensions and one time dimension), suggesting a total of eleven dimensions.

Q: What are some key concepts in M Theory?

A: Some key concepts in M Theory include the existence of extra dimensions beyond the four we experience, the idea of vibrating strings and higher-dimensional objects known as "branes," and the possibility of a multiverse consisting of multiple parallel universes with different physical properties.

Q: What are some implications of M Theory?

A: M Theory could potentially provide a unified framework for understanding all fundamental forces of nature, including gravity, electromagnetism, and the strong and weak nuclear forces. It also suggests the existence of multiple universes with varying physical laws, which could explain the apparent fine-tuning of our universe for life.

Q: What are some challenges facing M Theory?

A: One major challenge is the mathematical complexity of M Theory, which makes it difficult to test and verify experimentally. Additionally, there is currently no direct experimental evidence for the existence of extra dimensions or other predictions of

M Theory, leading to some skepticism within the scientific community. Despite these challenges, researchers continue to refine the theory and explore new approaches to testing its predictions.

Q: What is cryonics?

A: Cryonics is the practice of preserving human bodies or brains at very low temperatures with the hope of reviving them in the future, when medical technology may have advanced enough to restore health and vitality.

Q: How does cryopreservation work?

A: Cryopreservation involves rapidly cooling biological tissues to extremely low temperatures, typically using cryoprotectants to prevent ice crystal formation and cellular damage. The tissues are then stored in specialized facilities at temperatures around -196°C (-321°F).

Q: What are some of the challenges associated with cryonics?

A: One of the main challenges is achieving successful revival after cryopreservation. Current techniques may cause cellular damage during freezing and thawing processes, and the complexity of biological systems presents significant hurdles. Additionally, ethical concerns, such as informed consent and the long-term viability of storage facilities, need to be addressed.

Q: Are there any promising advancements in cryonics research?

A: Yes, advancements in fields like regenerative medicine and nanotechnology offer potential solutions for repairing cellular damage and restoring function in cryopreserved tissues. Ongoing research aims to refine cryopreservation techniques and improve our understanding of cellular preservation.

Q: What are some potential applications of cryogenics beyond human preservation?

A: Cryogenics has applications in preserving tissues and organs for transplantation, conserving genetic material for endangered species, and developing advanced bioprinting techniques for tissue engineering.

Q: What are some speculative scenarios for the activation of a time machine?

A: There are various imaginative possibilities, including a gradual acceleration through time from the Big Bang onwards, glitches and malfunctions causing chaotic jumps through random points in time, precise control allowing travel to specific destinations, sensory experiences transporting consciousness through time, and a blend of unpredictable outcomes with each activation.

Q: Are time machines currently possible?

A: No, time machines remain purely theoretical constructs based on scientific principles such as General Relativity and quantum mechanics. While they are a popular concept in science fiction, there is currently no scientific evidence or technology capable of enabling time travel.

Q: What does the future hold for cryonics and time travel?

A: The future of cryonics depends on continued advancements in medical technology and our understanding of cellular preservation. As for time travel, while it remains speculative, ongoing scientific research may one day uncover new insights into the nature of time and space, potentially leading to breakthroughs in theoretical physics.

GLOSSARY

1. M Theory: A theoretical framework in physics that aims to unify various fundamental forces and particles, including gravity, electromagnetism, and the strong and weak nuclear forces, by positing the existence of eleven dimensions.

2. String Theory: A theoretical framework that describes fundamental particles as one-dimensional "strings" vibrating at different frequencies, which may offer a path toward unifying quantum mechanics and general relativity.

3. Branes: Higher-dimensional objects in M Theory that vibrate within the eleven-dimensional space, potentially explaining the existence of multiple universes or "parallel branes."

4. Multiverse: The concept that there may exist multiple universes, each with its own set of physical laws and constants, possibly encompassing an infinite range of possibilities beyond our observable universe.

5. Vacua: Plural of "vacuum," referring to the potential states of lowest energy in a given physical system, which in the context of M Theory may correspond to different universes or dimensions.

6. Parallel Universes: Hypothetical universes that coexist alongside our own, potentially separated by different dimensions or spatial coordinates, as theorized in various interpretations of quantum mechanics and cosmology.

7. White Holes: Theoretical celestial objects speculated to be the counterparts of black holes, where matter and energy are emitted outward instead of being pulled inward due to extreme gravitational forces.

8. Cryogenics: The practice of preserving biological materials at very low temperatures, often for medical or scientific purposes, such as the preservation of tissues, organs, or even entire organisms for future revival or study.

9. Vitrification: A cryopreservation technique that involves rapidly cooling biological materials to prevent the formation of ice crystals, thereby reducing cellular damage during freezing.

10. Temporal Paradoxes: Hypothetical situations in which the existence of time travel leads to logical contradictions or inconsistencies, such as the grandfather paradox or the bootstrap paradox.

11. Event Horizon: The boundary surrounding a black hole beyond which nothing, not even light, can escape due to the extreme gravitational pull.

12. Singularity: A point within a black hole where matter is infinitely dense and space-time curvature is infinite, marking the breakdown of classical physics and our current understanding of the laws of physics.

13. Wormholes: Hypothetical tunnels or shortcuts through space-time that could connect distant points in the universe, potentially allowing for faster-than-light travel or even time travel.

14. Regenerative Medicine: A field of biomedical research focused on developing therapies that stimulate the body's own regenerative capabilities to repair or replace damaged tissues or organs.

15. Nanotechnology: The study and application of extremely small structures, typically at the scale of individual atoms or

molecules, with potential applications in medicine, materials science, and electronics, among others.

SUGGESTED READING

1. "The Elegant Universe: Superstrings, Hidden Dimensions, and the Quest for the Ultimate Theory" by Brian Greene - Provides a comprehensive introduction to String Theory and its implications, offering insights into the fundamental nature of reality.

2. "Warped Passages: Unraveling the Mysteries of the Universe's Hidden Dimensions" by Lisa Randall - Explores the concept of extra dimensions beyond the familiar four and their significance in modern physics, offering a deeper understanding of M Theory.

3. "Our Mathematical Universe: My Quest for the Ultimate Nature of Reality" by Max Tegmark - Examines the philosophical and scientific implications of the multiverse concept, delving into the idea of multiple universes with varying physical laws.

4. "Lost in Math: How Beauty Leads Physics Astray" by Sabine Hossenfelder - Provides a critical analysis of the challenges facing modern theoretical physics, offering insights into the current state of theoretical research and the quest for a unified theory.

5. "The Fabric of the Cosmos: Space, Time, and the Texture of Reality" by Brian Greene - Explores the nature of space, time, and the universe, discussing concepts such as relativity, quantum mechanics, and the search for a unified theory of physics.

6. "Parallel Worlds: A Journey through Creation, Higher Dimensions, and the Future of the Cosmos" by Michio Kaku - Investigates the concept of parallel universes and their

implications for our understanding of the cosmos, providing a fascinating exploration of theoretical physics.

7. "The Hidden Reality: Parallel Universes and the Deep Laws of the Cosmos" by Brian Greene - Explores the concept of parallel universes from a scientific perspective, discussing the various theories and evidence supporting the existence of multiple realities.

8. "The Grand Design" by Stephen Hawking and Leonard Mlodinow - Discusses the latest developments in theoretical physics, including M Theory, the multiverse, and the search for a final theory of everything, offering insights into the nature of reality.

9. "Quantum Space: Loop Quantum Gravity and the Search for the Structure of Space, Time, and the Universe" by Jim Baggott - Explores alternative theories of quantum gravity, such as loop quantum gravity, and their implications for our understanding of the universe's fundamental structure.

10. "The Trouble with Physics: The Rise of String Theory, the Fall of a Science, and What Comes Next" by Lee Smolin - Critically examines the dominance of String Theory in theoretical physics and its impact on the field, offering perspectives on the future direction of research.

BOOKS BY THIS AUTHOR

Quantum Consciousness: How The Principles Of Quantum Mechanics Explain The Mysteries Of Human Experience

Journey into the enigmatic realm where physics and the human mind intertwine! Quantum Consciousness unlocks the secrets of our most profound experiences, weaving together the principles of quantum mechanics with the enigmatic tapestry of consciousness. In this captivating read, you'll explore the mind-bending implications of quantum physics on our perceptions, emotions, and the very nature of reality. Discover the uncanny parallels between quantum phenomena and the mysteries of the human psyche, unraveling the enigmatic connection between our inner world and the fundamental laws governing the universe. Through thought-provoking experiments and cutting-edge research, this book will challenge your understanding of reality and leave you questioning the boundaries between the quantum and the conscious. Embark on a transformative voyage, where the mysteries of human experience collide with the wonders of quantum mechanics, forever altering your perception of the world and the nature of consciousness itself!

The Ecology Of The Mind: Exploring The Relationship Between Inner And Outer Worlds

Embark on a transformative journey to understand the profound connection between our inner and outer worlds with "The

Ecology of the Mind." This captivating book delves into the intricate tapestry of our minds, exploring how our thoughts, emotions, and experiences shape our relationship with nature and the wider cosmos.

Through a captivating narrative that weaves together scientific insights, evocative anecdotes, and ancient wisdom, "The Ecology of the Mind" reveals how our inner landscape mirrors the ecosystems we inhabit. It unveils the profound impact our thoughts and actions have on the health of our bodies, communities, and planet.

Discover how our minds are constantly interacting with the environment, exchanging energy, information, and meaning. Explore the concept of "ecopsychology," which examines the deep connection between our psychological well-being and the natural world. Uncover the therapeutic power of nature and learn how spending time in green spaces can restore our mental and emotional balance.

This thought-provoking book challenges us to rethink the traditional boundaries between mind and nature, inviting us to see ourselves as integral parts of a vast and interconnected web of life. With its profound insights and practical guidance, "The Ecology of the Mind" empowers us to live more harmoniously with ourselves, each other, and the planet we call home.